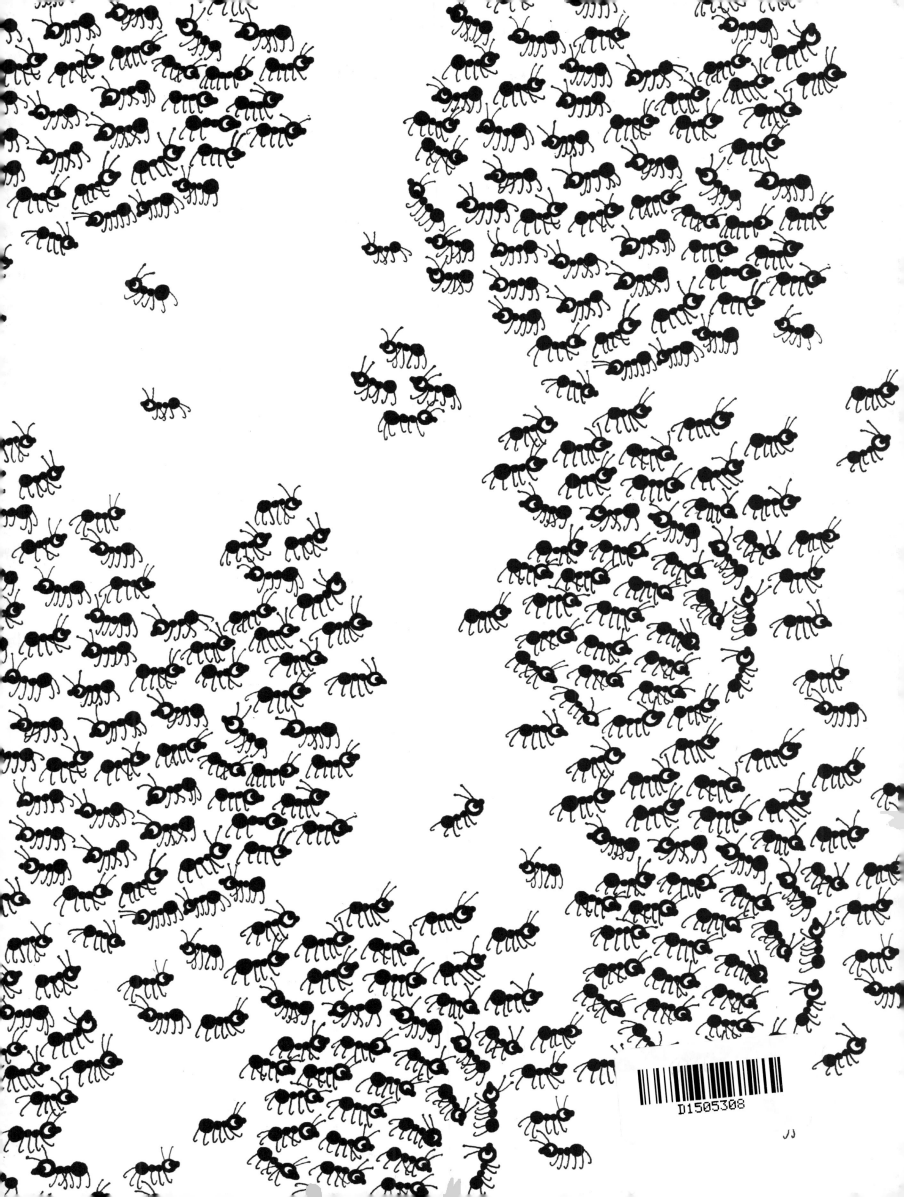

Translated from the French *Mille et une fourmis*

First published in the United Kingdom in 2019 by
Thames & Hudson Ltd, 181A High Holborn, London WC1V 7QX

www.thamesandhudson.com

First published in the United States of America in 2019 by
Thames & Hudson Inc., 500 Fifth Avenue, New York, New York 10110

www.thamesandhudsonusa.com

Original edition © 2018 Actes Sud, Arles
This edition © 2019 Thames & Hudson Ltd, London

British Library Cataloguing-in-Publication Data.
A catalogue record for this book is available from
the British Library

Library of Congress Control Number 2019934249

ISBN 978-0-500-65208-4

Printed in China

Joanna Rzezak

1001 ANTS

Thames & Hudson

In the middle of the forest is a strange little hill, covered with pine needles and sand. In fact, it's a home for thousands of ants. And all sorts of things are going on inside...

KEEP AN EYE OUT!
THERE'S A LITTLE ANT WITH RED SOCKS HIDING IN EVERY PICTURE IN THIS BOOK. CAN YOU FIND HER?

Worker ants come in different shapes and sizes. The smallest ones stay indoors. The biggest ones guard the nest. The medium-sized ones go out looking for food.

Little green insects called aphids make honey that ants love to eat. So the ants keep them in a mini aphid farm.

First, each egg will hatch into a little white wormy thing called a larva. Most of them are female and will grow up to be worker ants. But some will grow into male ants, and a few into new queens!

The egg room. These eggs will grow into new ants, but not straight away.

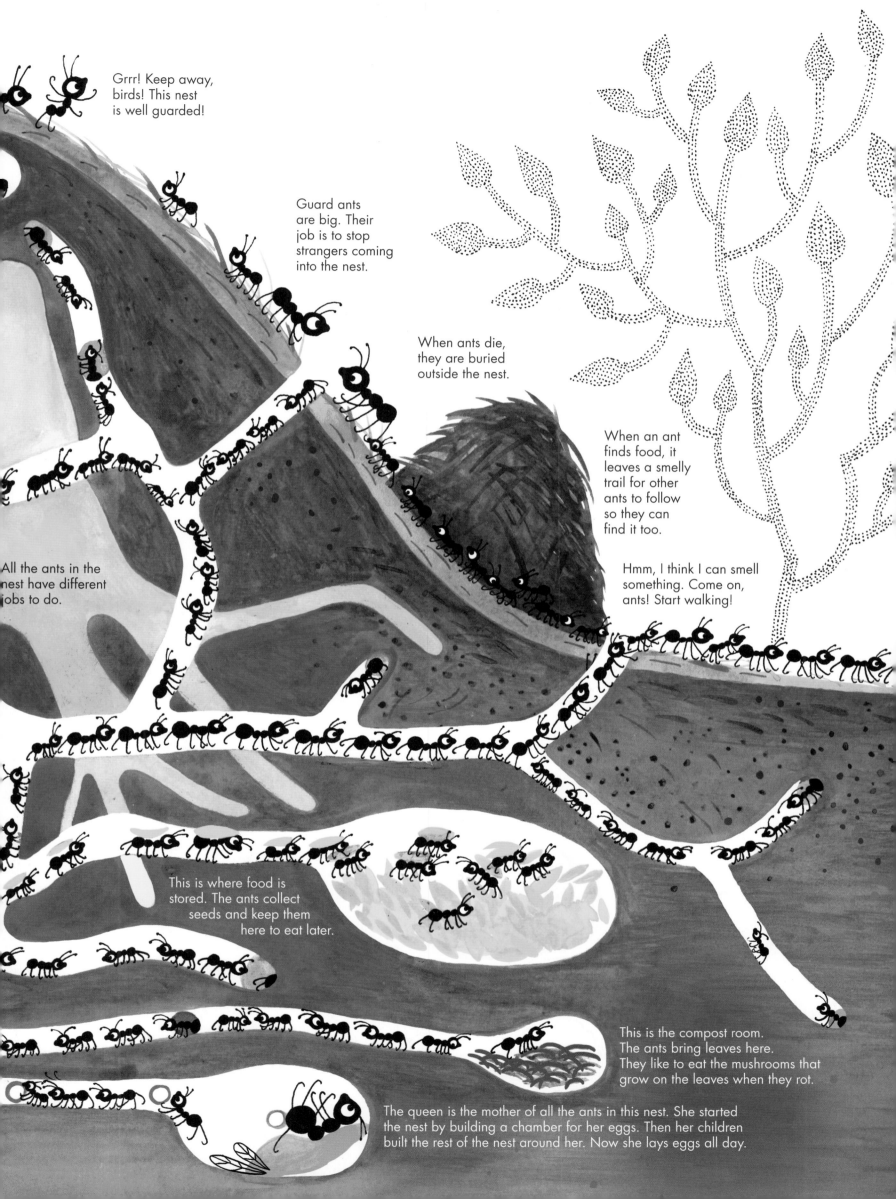

Grrr! Keep away, birds! This nest is well guarded!

Guard ants are big. Their job is to stop strangers coming into the nest.

When ants die, they are buried outside the nest.

When an ant finds food, it leaves a smelly trail for other ants to follow so they can find it too.

All the ants in the nest have different jobs to do.

Hmm, I think I can smell something. Come on, ants! Start walking!

This is where food is stored. The ants collect seeds and keep them here to eat later.

This is the compost room. The ants bring leaves here. They like to eat the mushrooms that grow on the leaves when they rot.

The queen is the mother of all the ants in this nest. She started the nest by building a chamber for her eggs. Then her children built the rest of the nest around her. Now she lays eggs all day.

Moss usually grows in damp places.

Hey, wait for Red Socks!

Lily of the valley is a pretty flower but watch out! It's poisonous.

Ants travel by walking single file in a long line. Keep going, ants! There's still a long way to go...

Ferns have been
growing on the Earth
for millions of years.
They are older than
the dinosaurs!

The Roman snail lives on
a patch of ground about
20 feet across. It spends its
whole life crawling around
and around inside it.

How many different types of
mushrooms can you recognize
in the wild?

Remember, some
mushrooms are toxic,
so look, but don't touch!
And never eat mushrooms
you find growing outside.

The cep is a mushroom
with a round cap and
a thick stalk.

The
morel
has an
unusual
shape,
like a little
knobbly
tree.

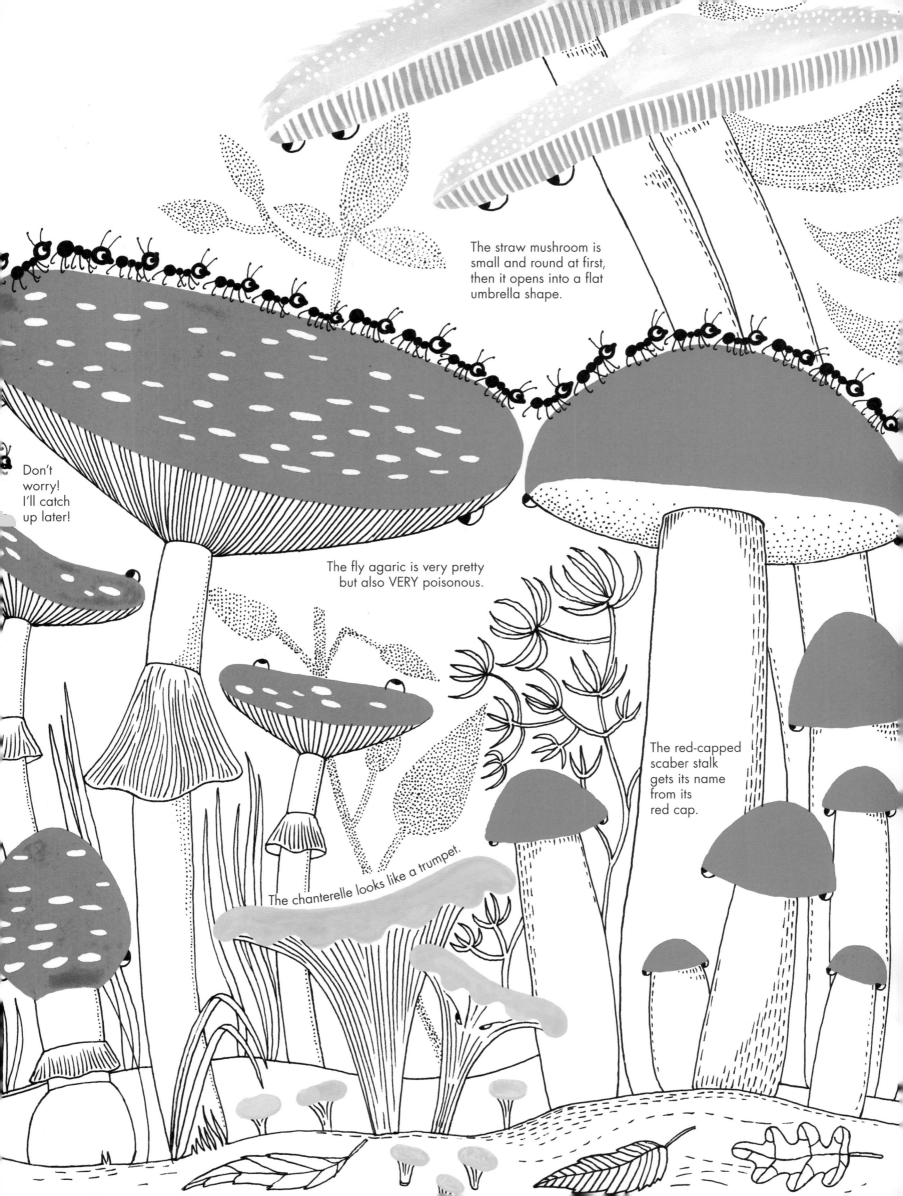

The straw mushroom is small and round at first, then it opens into a flat umbrella shape.

Don't worry! I'll catch up later!

The fly agaric is very pretty but also VERY poisonous.

The red-capped scaber stalk gets its name from its red cap.

The chanterelle looks like a trumpet.

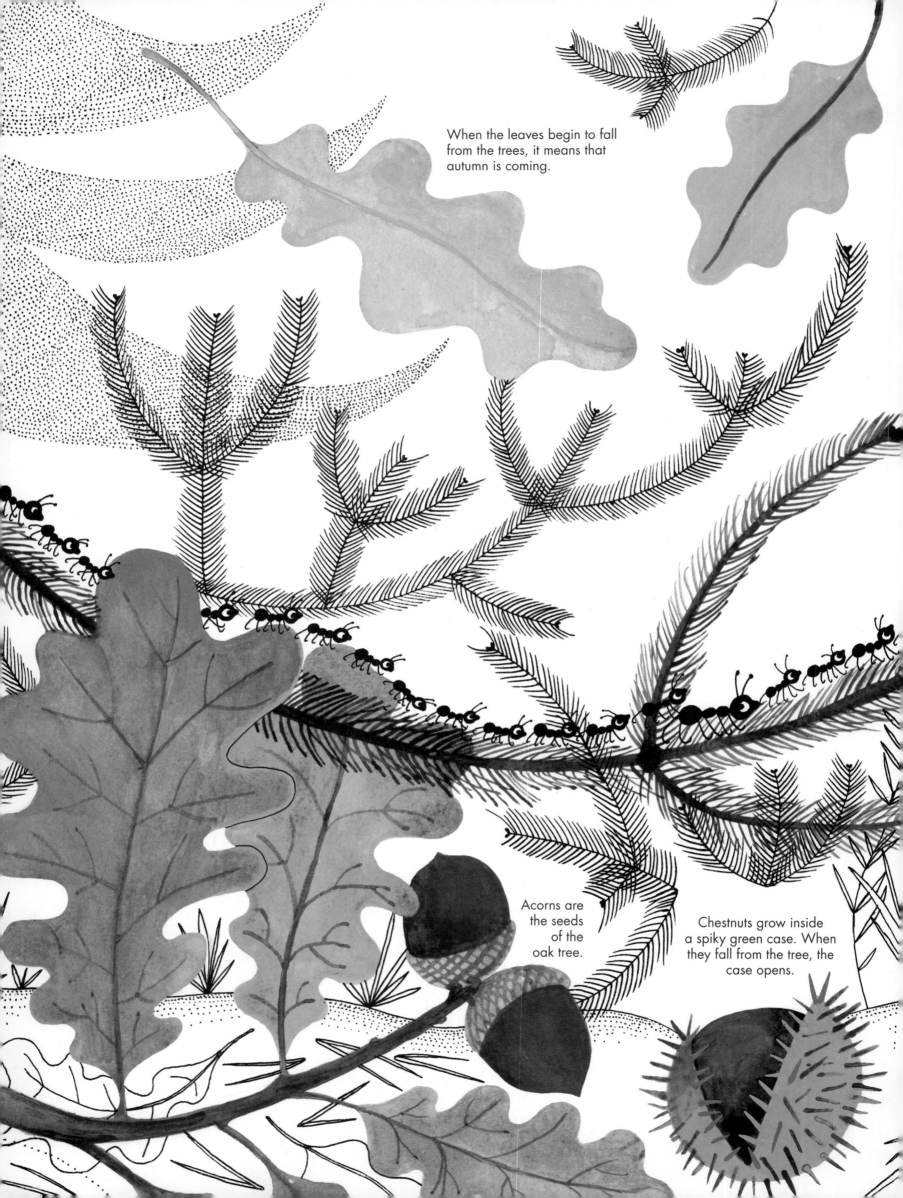

When the leaves begin to fall from the trees, it means that autumn is coming.

Acorns are the seeds of the oak tree.

Chestnuts grow inside a spiky green case. When they fall from the tree, the case opens.

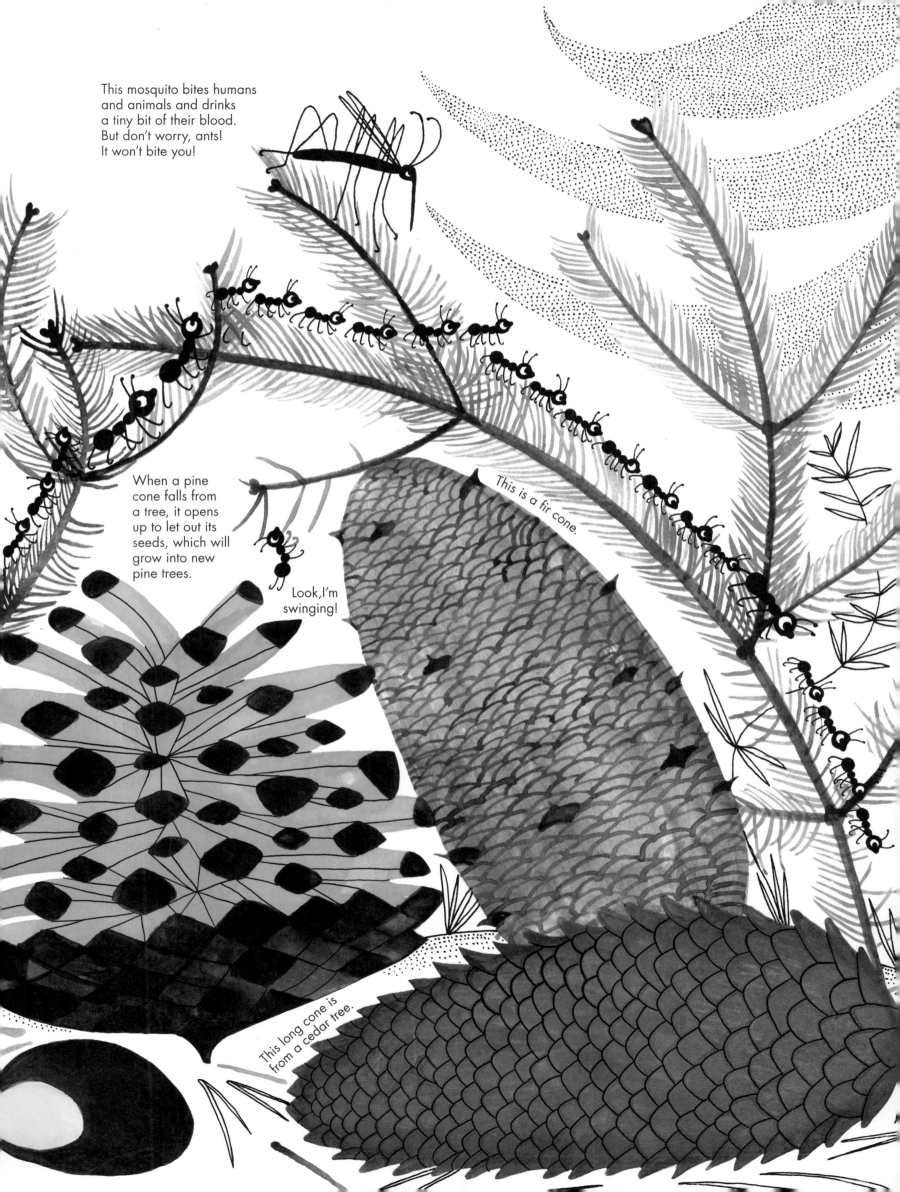

This mosquito bites humans and animals and drinks a tiny bit of their blood. But don't worry, ants! It won't bite you!

When a pine cone falls from a tree, it opens up to let out its seeds, which will grow into new pine trees.

Look, I'm swinging!

This is a fir cone.

This long cone is from a cedar tree.

The dragonfly can fly faster than any other insect. It can even catch mosquitoes in mid-air.

Don't stop here, ants! I'm sure these frogs would love to eat you.

Lily pads make a good place for animals to rest.

Reeds are tough and bendy plants that like to grow near water.

The flowers of the yellow water lily grow under the water, but they open when they reach the surface.

This insect is a pond skater. It's so light that it can walk on water.

The fluffy seeds of the dandelion are blown away by the wind.

This green grasshopper can jump more than 20 feet. That's 300 times its own size!

Ribwort is a very useful plant. You can rub the leaves on insect bites or nettle stings to make them feel better.

This is a
wheat
stalk.

This is an
oat stalk.

Nice
to meet
you!

This caterpillar
will grow into
a swallowtail
butterfly. It makes
a bad smell to
scare predators
away. Hold your
nose, Red Socks!

Beware of ticks! They can bite you.

A spider's web is a trap for insects. They get caught in the sticky threads and can't escape.

The male spider is usually smaller than the female. He's afraid she might eat him, so he often brings her a dead insect to eat instead!

It's not my lucky day...

The dung beetle collects animal droppings as food. It rolls them along the ground in a big ball. It's an expert at recycling!

The cross spider gets its name from the cross shape on its back. It can spin very big webs.

The female spider sits in the corner of her web and waits. When an insect lands on the web, she can feel the threads moving.

The spider weaves its web with a thread that comes out of a special hole at the tip of its tummy. The thread is very thin but very strong.

Is this a big furry hill? No, it's a sleeping bear! In the winter, bears can sleep for up to five months.

The caterpillar has turned into a swallowtail butterfly!

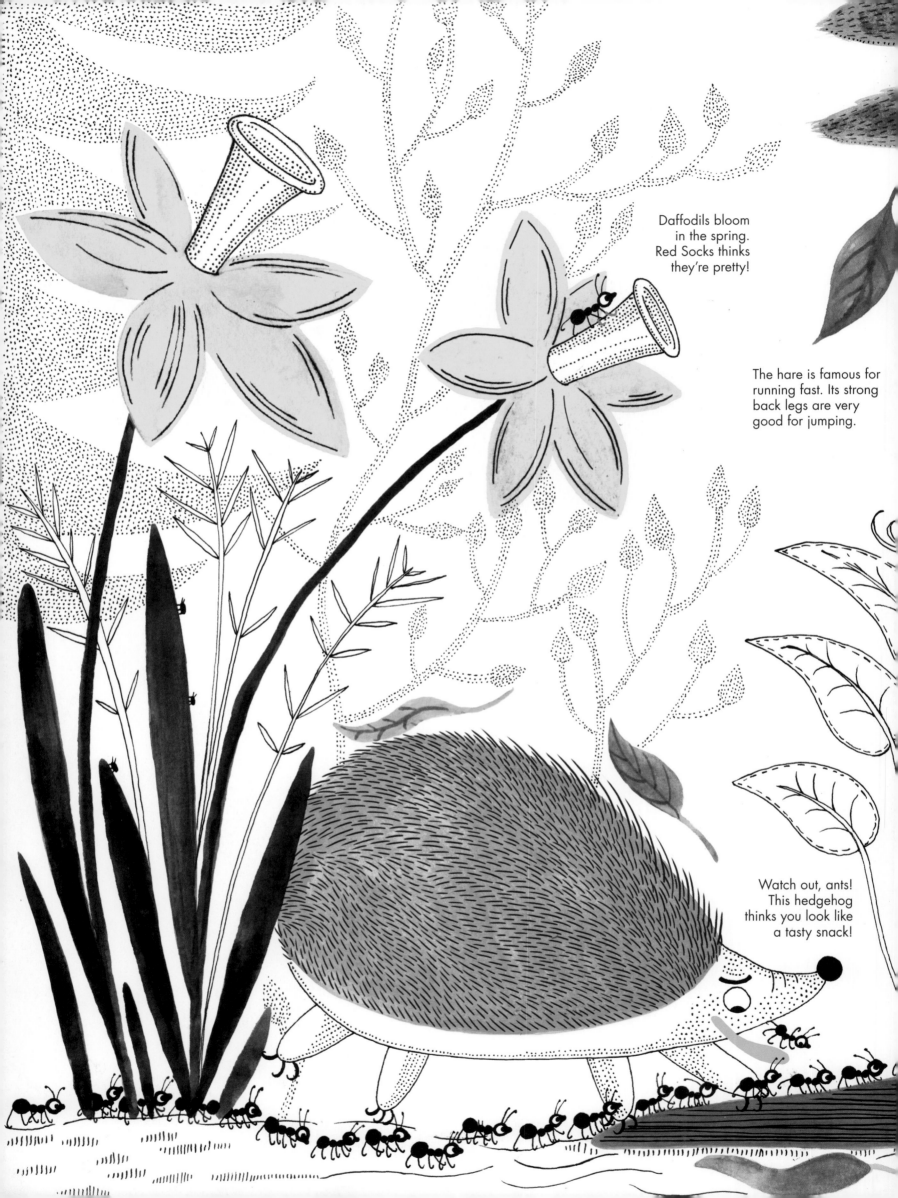

Daffodils bloom in the spring. Red Socks thinks they're pretty!

The hare is famous for running fast. Its strong back legs are very good for jumping.

Watch out, ants! This hedgehog thinks you look like a tasty snack!

Centipedes climb trees to find a sunny spot to rest.

Moss usually grows on the shadiest side of a tree.

Honey fungus is a kind of mushroom that often grows around the bottom of trees.

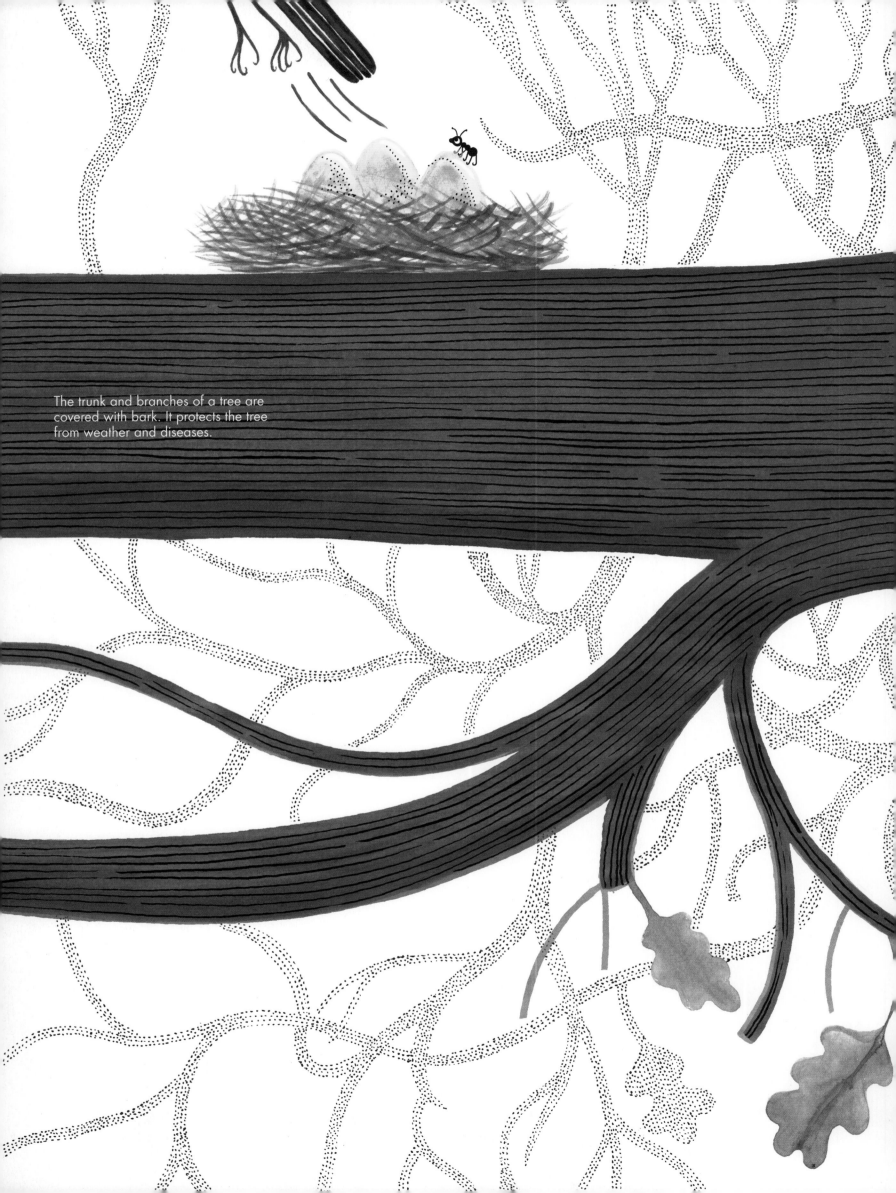

The trunk and branches of a tree are covered with bark. It protects the tree from weather and diseases.

Lichen is a kind of fungus. When you see it growing on a tree, it means that the air quality nearby is good.

These beetles are called wood-borers. Their grubs love to eat wood and can be dangerous to trees.

Whoops! An ant has fallen! But don't worry. It's very light, so it won't hurt itself when it lands.

Most moths sleep during the day and come out at night. The dots on their wings look like eyes to scare predators away.

Owls come out to hunt when it's dark. They talk to other owls by hooting.

The acorns are ready to fall from the tree. A new oak tree will grow wherever they land.

Well, hello there!

The wood of the oak tree is very strong. It's often used for building houses and making furniture.

The seeds of the sycamore tree have wings.

They spin through the air when the wind blows.

Hey, what's that noise? Bang! Bang! Bang! It sounds like a drum.

Come on, ants! Keep marching! You're just in time...

Snails can only move forward. Their average speed is three inches a minute.

...for my dinner.

Yum! We woodpeckers just LOVE to eat ants.

But not me. Yay!

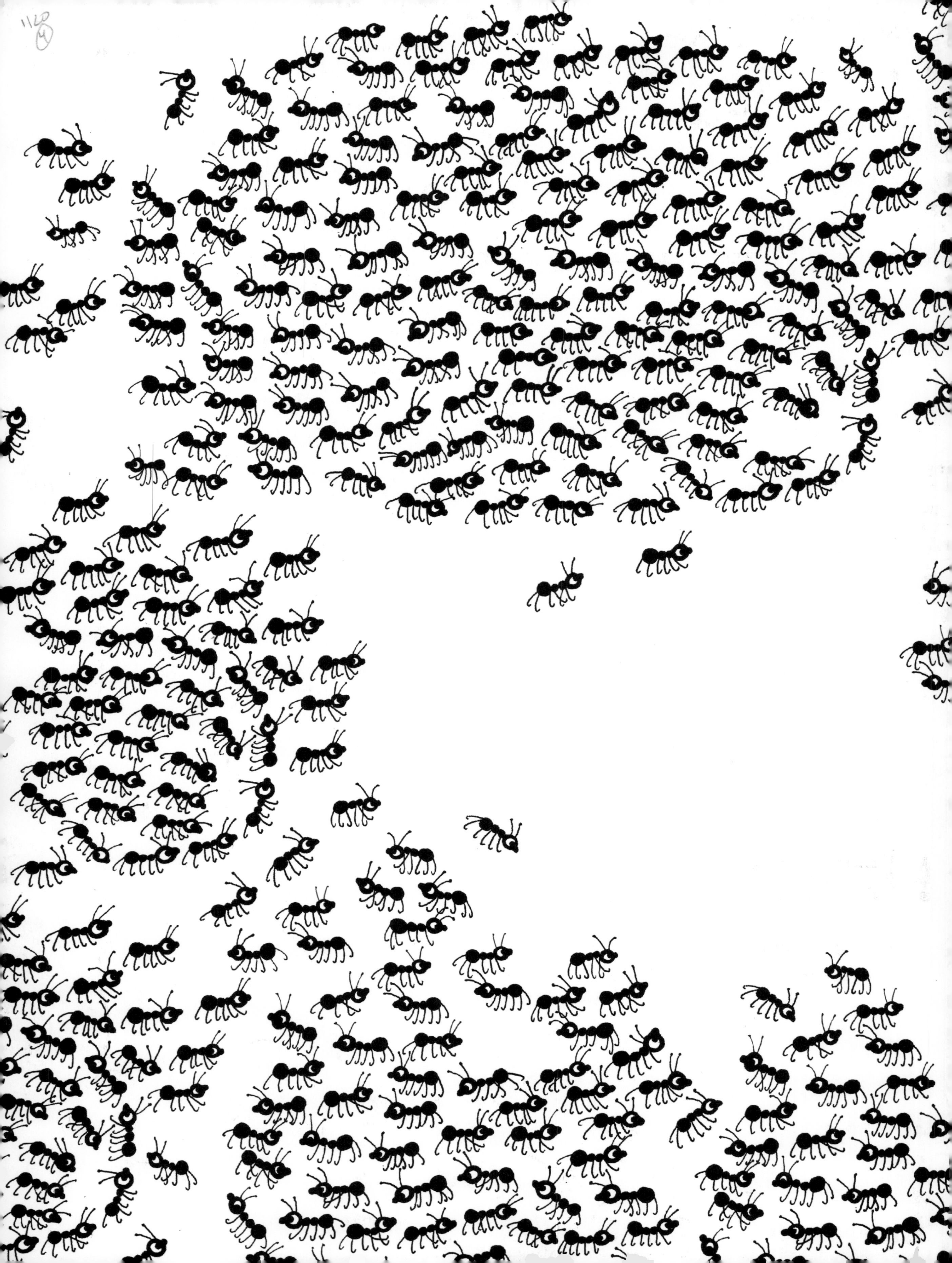